The Structure of Earth

by Laura McDonald

TABLE OF Contents

Introduction

Chapter 1

Chapter 2

Chapter 3

Conclusion

What have scientists learned about the structure of Earth?

Drilling to the Mantle

You can find life on the surface of Earth. But what can you find below the surface, inside Earth?

A long time ago, some people thought that Earth was an empty ball. Other people thought it had life inside it. In the book *A Journey to the Center of the Earth*, people see plants, mushrooms, and animals under the surface. But that was only a story.

The real-life Earth is huge. People live on Earth's surface, or crust. Below the crust, scientists think you will find a hot layer of rock called the mantle. Below that layer is an even hotter core. No person has seen the mantle or core. We learn about the inside of Earth in different ways. Earthquakes and volcanoes can teach us things. Rocks from space also teach us about Earth. Scientists use what they learn to make a model of the layers of Earth. The model changes as scientists learn more about Earth's structure.

One team of scientists wanted to learn more about the inside of Earth. This team wanted to drill a hole through Earth's crust. The Kola Superdeep Borehole would drill to where they thought the mantle began.

Drilling began in 1970. As they drilled, scientists studied samples from the hole. The samples surprised scientists. First, they found minerals they did not expect to find. Second, they found fossils much deeper than they expected. Third, they found rocks that had water inside. No one could explain where the water came from. Finally, they found that the temperature inside Earth was much higher than they had thought it would be.

By 1979, the hole was almost 10,000 meters (6 miles) deep. What would the scientists find when they went even deeper? Read this book to find out!

▼ Jules Verne imagined that there were ancient monsters living deep inside Earth.

▼ The Kola Superdeep Borehole in Russia was drilled deep into Earth's crust to bring up rock samples.

Barents Sea

Kola Superdeep Borehole

Murmansk

White Sea

Archangelsk

SWEDEN

FINLAND

RUSSIA

St. Petersburg

Baltic Sea

ESTONIA

LATVIA

Moscow

LITHUANIA

Studying Earth's Interior

How do scientists learn about the structure of Earth?

Look at the photo. This is a view of Earth from space. Earth looks calm. We see wide blue oceans over most of the planet's surface. But what lies beneath Earth's thin shell? Scientists believe that deep underground lies molten, or melted, rock. They also believe that Earth's center is a super-hot **core** (KOR).

Why do scientists think this? Sometimes volcanoes erupt. Other times earthquakes shake the ground. These events give us clues about what lies inside Earth. Scientists have worked for hundreds of years. They have collected evidence from many sources. Each new piece of evidence is like a puzzle piece. Each one helps Earth scientists understand the structure of Earth.

While we know a lot about Earth's outer surface, the layers of Earth's interior remain a puzzle that scientists are still working to solve. ▶

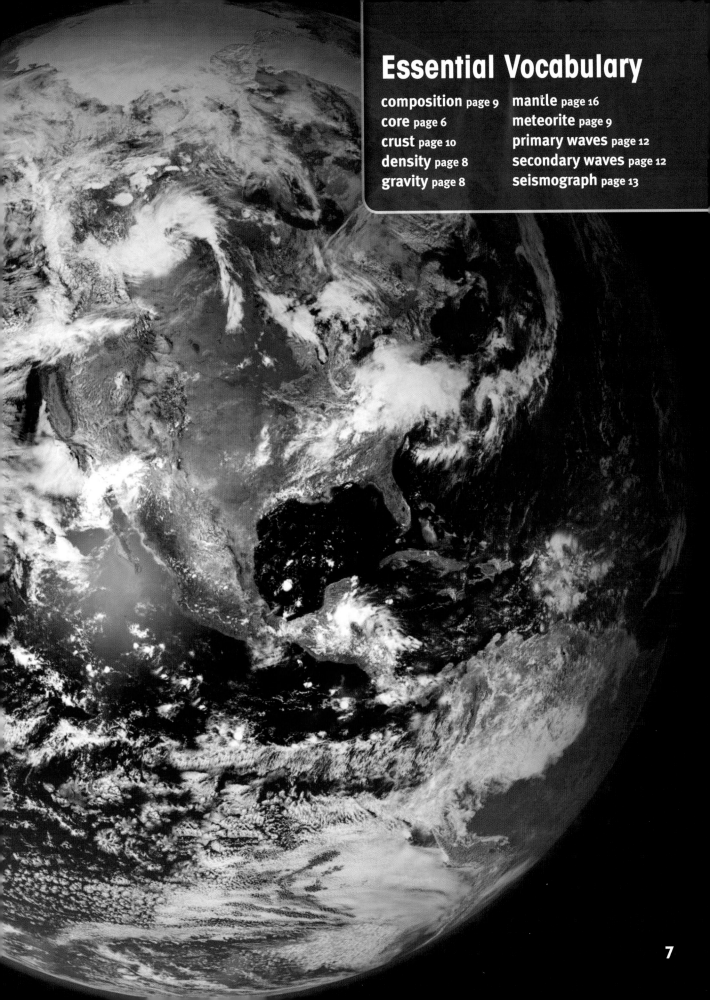

Essential Vocabulary

Evidence from Earth

Long ago, people thought that Earth was a flat disk floating on an endless ocean. Other people thought that Earth was a mountain or a hollow ball.

Aristotle (A-rih-stah-tul) was a Greek thinker. He was also an astronomer (uh-STRAH-nuh-mer). He studied the stars more than 2,000 years ago. Aristotle saw that Earth cast a round shadow on the moon. Aristotle's ideas led to the idea that Earth is a sphere, or round ball.

Henry Cavendish was an English scientist. Cavendish showed that Earth must have layers. Earlier scientists had found **gravity** (GRA-vih-tee). This is the force of attraction between all objects due to their mass. The greater an object's mass, the more gravity it has. Cavendish used Earth's gravity to find the mass of the planet. He compared the mass and size of Earth. Then he found Earth's **density** (DEN-sih-tee). Density is a measure of how much mass is in a given space, or volume (VAHL-yoom).

Cavendish showed that Earth's density was higher than the density of surface rocks. This meant denser layers must be under the ground. Today, scientists agree. Scientists think that Earth has a solid inner core with lighter layers around it.

Geologists (jee-AH-luh-jists) are scientists. Geologists study the structure and processes of Earth. Scientists cannot see the layers inside Earth. But geologists can study space rocks that fall to Earth. They can also look at materials that rise up from inside Earth.

Careful observations showed Aristotle that Earth casts a circular shadow on the moon. ▼

Math & Science
Units of Density

Density helps scientists predict how liquids form layers. For example, most oils float on water because oil is less dense than water. Density is described in terms of mass per unit volume, or m/v. For example, the density of liquid water is 1 gram/milliliter (8.3 pounds/gallon). The density of Earth is 5.5 grams/cubic cm (46 pounds/gallon).

The ROOT
of the Meaning

Meteor comes from the Greek roots *meta-* ("high up") + *-aoros* ("lifted, hovering in air"). A meteoroid is a rock that falls through space, a meteor is a rock that enters our atmosphere, and a meteorite is a rock that hits Earth's surface.

Scientists learn about Earth's structure from meteorites, which fall to Earth's surface. ▼

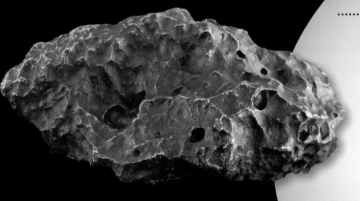

Evidence from Space

Have you ever seen a shooting star? Shooting stars are really light given off by small pieces of rock falling from space. We call these rocks meteoroids. Meteoroids can burn up as they fall through Earth's atmosphere. Sometimes a piece of meteoroid survives the fall. We call a rock that lands on Earth a **meteorite** (MEE-tee-uh-rite).

The solar system formed about 4.6 billion years ago. Scientists have found meteorites from that time. They study the **composition** (kahm-puh-ZIH-shun), or makeup, of these rocks. Scientists have found that these rocks have the same types of minerals as Earth.

Meteorites give us clues about the layers of rock inside Earth. Some meteorites have about twenty-seven percent iron. Surface rocks on Earth have only five percent iron. That is why scientists think you will find a lot of iron inside Earth.

Science to Science
Astronomy and Geology

Large meteorites have the potential to destroy human property and lives. The National Aeronautics and Space Administration (NASA), the United States space agency, keeps track of all large space objects that are predicted to come near Earth in the next 100 years.

Volcanoes

Volcanoes show us that Earth's inner layers are different from the surface. Volcanoes erupt with molten rock called lava. Lava oozes, pours, or splashes from volcanoes. Lava can get as hot as 1,160°C (2,120°F). Hot ash, rock, and gas come out of volcanoes. This rock and ash can get as hot as 700°C (1,290°F). Nothing on Earth's surface gets as hot as a volcano.

Where does this hot stuff come from? Deep inside Earth are pools of melted rock, or magma. Under pressure, the magma rises. As the magma rises, its heat melts more rock. When the pressure gets too high, or if a crack forms, the magma pours out of the **crust** (KRUST). Volcanoes are vents that let hot materials escape from inside Earth.

Geologists learn about the structure of Earth by looking at volcanic materials. Volcanoes bring up rock from 100 or more kilometers inside Earth.

Careers in Science
Volcanologist

▼ Volcanic eruptions bring materials from the inner layers of Earth to the surface.

▲ A volcanologist is a scientist who studies how volcanoes work. Aspiring volcanologists usually study geology in college. They work both in the lab and in the field. Volcanologists may teach in a university or consult with governments about planning volcano safety.

lava

magma

▲ inside a volcano

◄ Geologists sample lava from
volcanoes to learn about
Earth's inner structure.

Evidence from Earthquakes

Earth's crust is made of different pieces, or plates. When the plates of Earth's crust collide, earthquakes happen. Earthquakes make waves of energy that go through Earth. Geologists study these waves. They use these waves to understand Earth's structure.

Earthquakes make different types of waves. **Primary waves** (PRY-mair-ee WAVEZ), or P waves, are the fastest. P waves get to any given point first. P waves travel through solids, liquids, and gases. P waves move faster through denser matter. P waves move back and forth like a spring.

Secondary waves (SEH-kun-dair-ee WAVEZ), or S waves, move up and down. S waves are like waves in water. S-waves can also move side to side like a snake. S waves are slower than P waves. S waves travel through solids. But they do not travel through liquids and gases. S waves speed up as they move through denser materials.

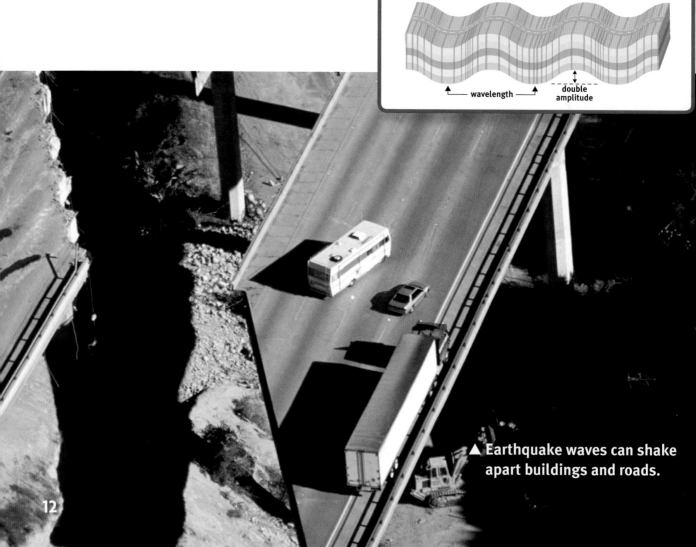

Earthquake waves can shake apart buildings and roads.

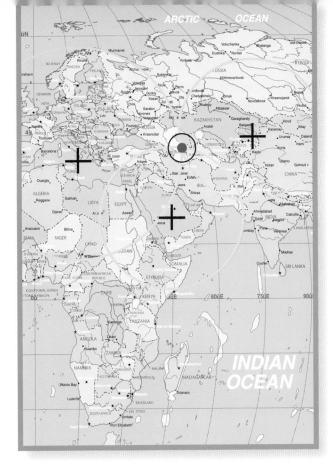

▲ **By drawing circles around at least three seismograph locations and seeing where the circles intersect, scientists can find the epicenter of the earthquake.**

Earthquake waves spread out in all directions from the center of the quake. Instruments called **seismographs** (SIZE-muh-grafs) tell geologists where the epicenter, or origin, of an earthquake is.

Around 1900, geologists placed seismographs all around the world. Geologists saw that earthquake waves sped up as they moved deeper than 200 kilometers (160 miles) into Earth. Geologists figured out that a denser layer must be under Earth's crust.

At a depth of 2,900 kilometers (1,800 miles), P waves slowed and S waves stopped. This clue tells us that Earth has a layer of liquid deep inside. The study of earthquakes tells us about Earth's inner layers.

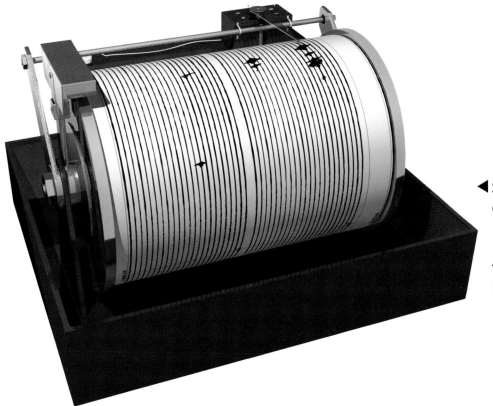

◄ **Seismographs record earthquake waves. Geologists study how these waves travel to better understand Earth's interior.**

Hands-On Science

Solid Waves

You've seen waves and ripples moving through water. Earthquakes generate waves through solid ground. How do waves move through a solid?

Try these waves!

- Make the spring shorter or longer and measure the time it takes a wave to go from one end to the other.

- Have the holder let go of the spring before you make waves.

- Have the holder and wave generator make waves at the same time.

- Make patterns of small and large waves.

Time Required
30–45 minutes

Materials Needed
- pencil and paper to record observations
- long spring
- stopwatch
- meter stick
- uncarpeted floor or pavement surface to work on

Procedure

1. Form a group of four students. Each person has a job. The jobs are: timer, spring holder, wave generator, and observer. Be sure to trade jobs during the activity so that each person gets to do each job.

2. The holder and wave generator should each take one end of the spring, move 2 meters (7 feet) apart, and then sit down and hold both ends of the spring on the floor.

3. First, the wave generator produces push waves by hitting the end of the spring. Then the wave generator produces up-and-down waves by jerking the spring once up and down. The observer draws a sketch of each type of wave.

4. The timer determines how long it takes each type of wave to travel down the length of the spring from the wave generator to the holder. The timer should time each type of wave three times and calculate the average time for push waves and up-and-down waves.

5. The wave generator can create some new types of waves with the spring. The observer should draw sketches of the different waves produced. Be careful not to tangle or overstretch the spring.

Hands-On Science

Analysis

1. Do push waves or up-and-down waves move faster?

2. Which type of wave represents primary or P waves? Which represents secondary or S waves? Explain your reasoning.

3. Describe some of your original wave experiments and their results.

The Current Model of Earth's Structure

Earthquakes show us that Earth is like an onion. Earth has many layers. At the center is the dense iron core. The inner part of the core is solid. The outer part of the core is liquid. Above the core lies the **mantle** (MAN-tul). The mantle has solid and partly melted layers. The thin outer layer of Earth is the crust. The crust is solid rock.

Earth's Interior

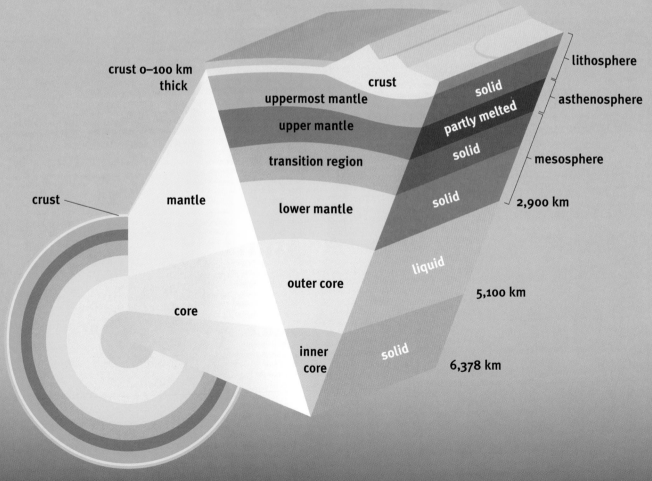

crust 0–100 km thick

crust

uppermost mantle

upper mantle

transition region

lower mantle

outer core

inner core

crust

mantle

core

lithosphere

asthenosphere

mesosphere

solid

partly melted

solid

solid

liquid

solid

2,900 km

5,100 km

6,378 km

(diagram not to scale)

Science and Technology
Shaker Trucks

Scientists can produce their own small earthquake-like tremors using shaker trucks. These massive trucks bounce up and down to produce waves in the ground. Geologists use seismographs to detect how the waves travel through Earth's crust.

▼ This shaker truck can produce earthquake-like tremors.

Earth's Early History

Scientists believe that the solar system began about 4.6 billion years ago. Earth and the rest of the solar system began as a cloud of gas in space. The gas particles cooled to form bits of dust. Gravity pulled these bits of dust together. Dust grains formed rocky lumps. These lumps bumped into one another and joined, getting bigger. As the rocks grew bigger, their gravity also grew. Rocks crashed together to make larger masses. One of these large masses was the beginning of Earth.

The greater Earth's mass grew, the stronger its gravity became. The young Earth went around the sun. It collected all of the dust and rocks in its path. Rocks that fell and hit Earth's surface became meteorites. Meteorites heated Earth so much that its rocks melted.

Earth had its biggest collision soon after the planet formed. A Mars-size planet about 6,400 kilometers (4,000 miles) wide crashed into Earth. The impact threw a huge spray of material into space. Some of this rock combined to form the moon.

In time, collisions decreased. Denser material sank toward the middle of Earth. This formed the core. Since the Earth rotates, the liquid outer core spins around the inner core. Heat went from the core to the mantle layer. Lighter materials rose through the mantle to Earth's surface. Then this layer cooled enough to form the solid crust. Gases bubbled up to form Earth's atmosphere. Earth then had the structure we study today.

▲ Astronomers think that Earth's moon was created by an impact between Earth and another small planet 4.6 billion years ago.

✔ CHECKPOINT

Visualize It

Meteorites of every size bombarded the early Earth. Draw how this might have appeared. Remember that there was no life yet on Earth.

umming Up

The only layer of Earth visible to people is the outer layer.

Scientists use indirect evidence from Earth measurements, meteorites, volcanoes, and earthquakes to learn about what's inside the planet.

This evidence reveals that Earth consists of several layers.

One theory states that Earth's layers formed during a violent early history of impacts from space rocks.

Putting It All Together

Choose one of the research activities below. Work independently, in pairs, or in small groups. Share your responses with your class.

1 Reread page 9 about meteorites. Use your library or the Internet to find out more about a particularly famous meteorite. Share your findings in a report to your class.

2 One of your friends tells you that Earth is a solid piece of rock. How would you respond? Write a short play describing the conversation between you and your friend.

3 Suppose that you received a wrapped gift and wanted to figure out what was inside without opening it. Describe seven tests you could perform on the gift to find out the contents without damaging the package. Explain how this activity relates to what geologists have learned about the structure of Earth.

CARTOONIST'S NOTEBOOK
ILLUSTRATED BY TIANYIN YANG

VOLCANOES AND EARTHQUAKES CAN TELL US A LOT ABOUT EARTH'S INNER STRUCTURE.

WHAT ARE SOME THINGS WE KNOW ABOUT EARTH AS A RESULT OF STUDYING VOLCANOES AND EARTHQUAKES?

WHAT ARE SOME THINGS YOU'D LIKE TO KNOW ABOUT EARTH'S STRUCTURE?

Earth's Core and Lower Mantle

What are the characteristics of Earth's core and lower mantle?

Earth is a huge, layered ball of rock. The core and lower mantle are the deepest layers. Each layer of Earth presses down on the one beneath it.

On Earth's surface, people feel only the pressure of the air above them. But the pressure inside Earth is much greater. The pressure squeezes and presses the rock deep inside Earth. This pressure makes the rock very dense and very hot. The deeper the layer, the greater the pressure. The core and lower mantle are the hottest, densest layers inside Earth.

▼ High temperature and pressure can change rocks from one type to another.

Everyday Science
Under Pressure

Essential Vocabulary

mesosphere page 28

When the air pressure outside your body is different from the air pressure inside your ears, you may feel discomfort. You may want to swallow to "pop" your ears. You could feel this type of pressure difference when flying in an airplane, driving in the mountains, or diving in a swimming pool. The pressure inside Earth can be up to three million times greater than the pressure on Earth's surface. We can measure this pressure in units called atmospheres (atm).

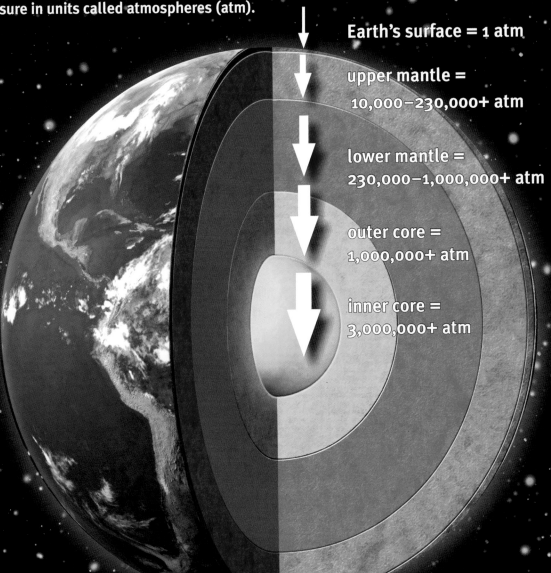

Earth's surface = 1 atm

upper mantle = 10,000–230,000+ atm

lower mantle = 230,000–1,000,000+ atm

outer core = 1,000,000+ atm

inner core = 3,000,000+ atm

The Core

The core is the innermost layer of Earth. The core is very, very hot. It is about the temperature of the sun's surface. Most of this heat comes from the breakdown of radioactive elements.

When it first formed, Earth was a mixture of all types of molten rock. Earth did not have a core, mantle, or crust. Over time, denser materials (like iron and nickel) sank and formed the core.

THE INNER CORE

Scientists believe that the whole core was once molten rock. Heat slowly rose upward toward Earth's outer layers. At the same time, the core also became less radioactive.

The core has been cooling for billions of years. The coolest parts become solids under the high pressure. Then they sink toward Earth's center. The solid inner core started to form two to four billion years ago. It has been slowly growing ever since.

THE OUTER CORE

The inner and outer core have a similar makeup. However, the inner core has a greater temperature than the outer core. The pressure in the outer core is also not as high. The pressure is not high enough to squeeze liquids into solids. Earthquake waves tell us that the outer core flows like a liquid.

Earth's inner core may be one gigantic crystal of iron and nickel. ▼

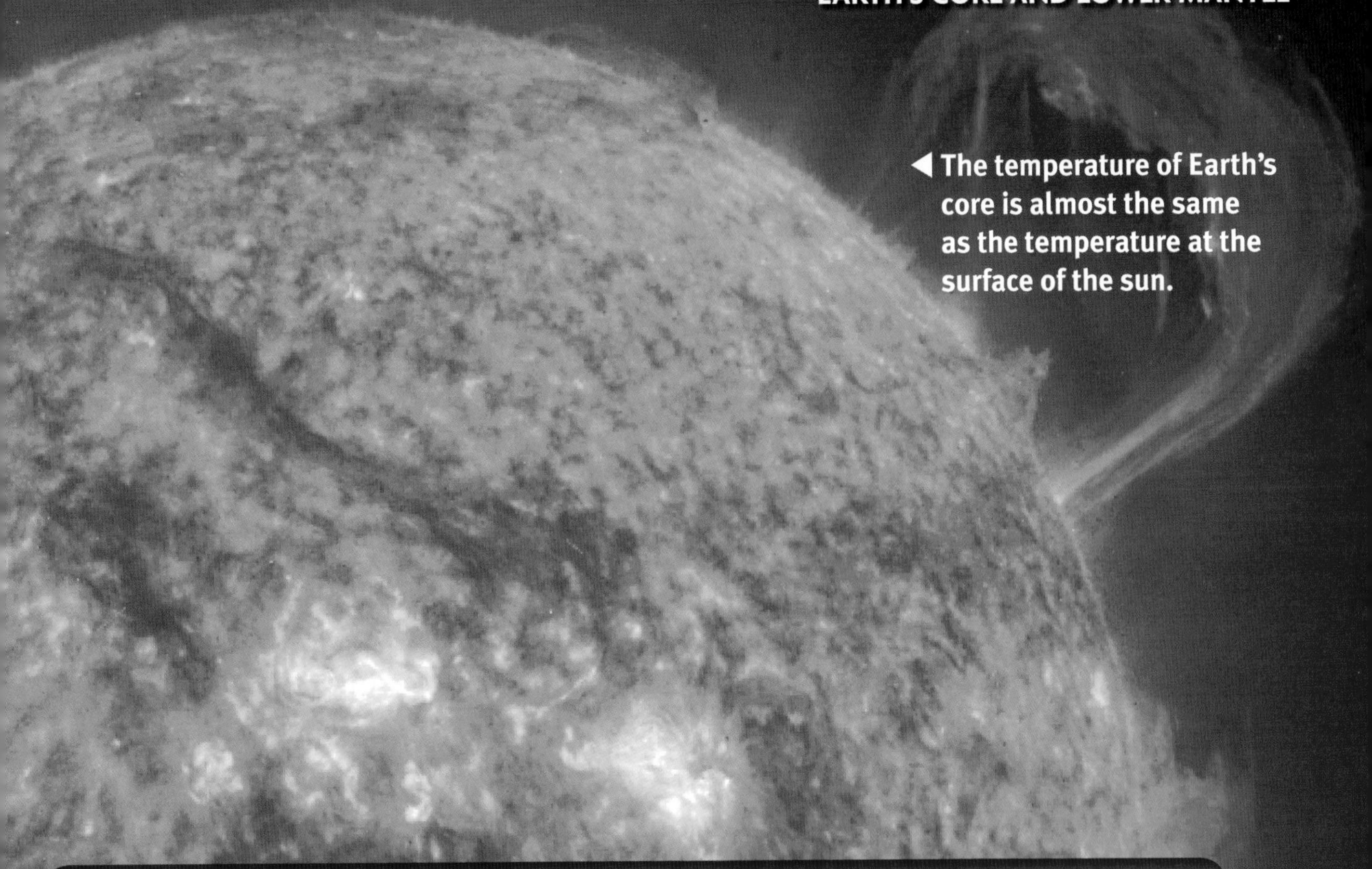

◄ The temperature of Earth's core is almost the same as the temperature at the surface of the sun.

Earth's Core

The Inner Core

Depth: 6,400–5,100 km (4,000–3,200 mi)

Radius: 1,300 km (800 mi)

Temperature: 5,000–6,000°C (9,000–11,000°F)

Pressure: Greater than 3 million atmospheres

State: Solid

Composition: Mostly iron and nickel

The Outer Core

Depth: 5,100–2,900 km (3,200–1,800 mi)

Thickness: 2,200 km (1,400 mi)

Temperature: 4,400–5,400°C (8,000–9,800°F)

Pressure: Greater than 1 million atmospheres

State: Liquid

Composition: Mostly iron and nickel, possibly some oxygen and sulfur

(measurements rounded to the nearest hundred)

Magnetic Fields

Currents in the outer core create a magnetic field. This magnetic field causes the whole planet to act like a giant magnet. The magnetic field helps living things in different ways. Sea turtles find their way home with the help of Earth's magnetic field.

The magnetic field also protects life on Earth. The magnetic field is like a shield. The magnetic field deflects the sun's dangerous particles. The particles that do hit Earth's atmosphere cause beautiful auroras in the sky. We can see these colorful displays in the far north and far south. Without a magnetic field, life could not exist on Earth.

✓ **CHECKPOINT**
Read More About It

The moon, Venus, Mercury, and Mars also have metallic cores at their centers. Read more about the structure of the other planets in our solar system on the Internet or in your school or local library.

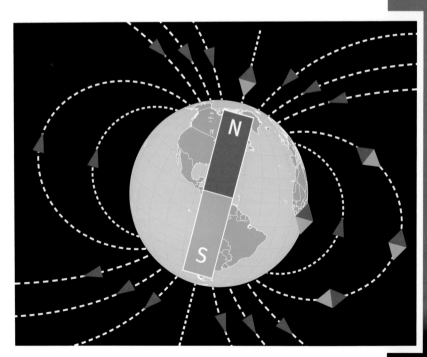

▲ The swirling of the outer core generates a magnetic field around Earth.

▲ Earth's magnetic field creates beautiful auroras.

They Made a Difference
Inge Lehmann (1888–1993)

Inge Lehmann (IN-guh LAY-mahn) was born in Europe at a time when few women were scientists. Lehmann was fortunate to attend a school where boys and girls studied the same subjects and were treated as equals.

▲ Inge Lehmann predicted the existence of Earth's solid inner core.

Lehmann had a special interest in the science of earthquakes. She kept records of many earthquakes at a scientific station she ran in Denmark. She compared her records with those made by other scientists at stations around the world. At that time, scientists believed that Earth had a liquid core surrounded by a solid mantle and crust.

In 1936, Lehmann observed the shock waves from a large earthquake in New Zealand. The timing of those waves puzzled her because they did not match the predictions of the current theories of Earth's structure. Based on her observations, Lehmann predicted the existence of a solid inner core surrounded by a liquid outer core.

Lehmann lived to the age of 105 and remained active in the field of geology well into her retirement.

The Lower Mantle

The lower mantle lies above the core. Some scientists call the lower mantle the **mesosphere** (MEH-zuh-sfeer). The mantle is made of oxygen, silicon, magnesium, and iron. These elements can combine to make silicates and other minerals. Geologists predict that we have never seen some of the minerals in the mantle on Earth's surface. Like the core, we cannot see or get samples from the lower mantle.

Almost half of Earth's mass is found in the lower mantle. Earthquake waves tell us that the lower mantle is solid. The temperature is hot enough to melt the lower mantle materials. But the pressure from the upper mantle and crust force the minerals back into the solid state.

Scientists know less about the lower mantle than any other part of Earth. More information about the lower mantle would help us understand how heat moves from the core to Earth's surface. The lower mantle may also play a role in the making of volcanoes.

The Lower Mantle

Depth:	2,900–660 km (1,800–410 mi)
Thickness:	2,240 km (1,390 mi)
Temperature:	2,000–4,000°C (3,600–7,200°F)
Pressure:	1,400,000 to 230,000 atmospheres
State:	Solid
Composition:	Mostly minerals containing oxygen, silicon, magnesium, and iron

(measurements rounded to the nearest ten)

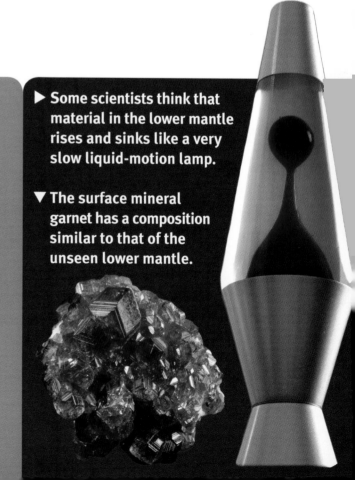

▶ Some scientists think that material in the lower mantle rises and sinks like a very slow liquid-motion lamp.

▼ The surface mineral garnet has a composition similar to that of the unseen lower mantle.

Careers in Science
Seismologist

Much of our understanding of Earth's deeper layers comes from the work of seismologists—scientists who study earthquakes and other ground vibrations. Seismologists study geology in college and often continue on to graduate school. They work for governments, universities, and companies that mine resources from underground.

Summing Up

- The innermost layers of Earth are the inner and outer core, and the lower mantle.
- These layers are extremely hot and compressed by the tremendous weight of the rock above.
- The inner core is made of solid iron and nickel.
- The outer core is a slowly swirling liquid that creates Earth's magnetic field.
- The lower mantle is made of solid minerals and remains largely a mystery to science.

Putting It All Together

Choose one of the research activities below. Work independently, in pairs, or in small groups. Share your responses with your class.

1 Earth's magnetic field protects life on Earth from dangerous radiation from space. Reread page 26 and then write a short science fiction story about a time in the future when Earth's outer core stops swirling and the magnetic field fades away.

2 Mix cornstarch and water in a pan. Try pressing down hard on the mixture with your finger. Compare your mixture to the rock under great pressure inside Earth. Discuss your findings with a friend.

3 Scientists are always conducting new studies about Earth's inner layers. Use the Internet to learn about the most recent discoveries related to the lower mantle and core. Share these discoveries in a short report to your class.

Earth's Upper Mantle and Crust

What are the characteristics of Earth's upper mantle and crust?

Unlike the core and lower mantle, scientists can study the upper mantle and crust. Lava from a volcano is evidence that molten rock lies beneath Earth's surface. Scientists can measure the temperature of the lava. They can study cooled lava under a microscope. They can also test the lava to find out what minerals it contains.

Scientists know a lot more about the upper mantle and crust than they do about Earth's inner layers. Evidence shows that heat rises from the lower mantle and core below. This heat forms currents in the upper mantle. The crust and top layer of the mantle float on this sea of hot rock. At times, pieces of crust bump or slide past each other. This creates earthquakes. Hot material can also break through the crust as volcanoes.

This volcanic chimney is made of mantle rock that broke through the crust many years ago. ▶

The Upper Mantle

Above the lower mantle lies the upper mantle. The upper mantle has two main layers. The **asthenosphere** (as-THEH-nuh-sfeer) makes up most of the upper mantle. The asthenosphere is a partially melted solid. The uppermost mantle is the top layer. This layer is solid rock.

The Root of the Meaning

Asthenosphere comes from the Greek word for "weak."

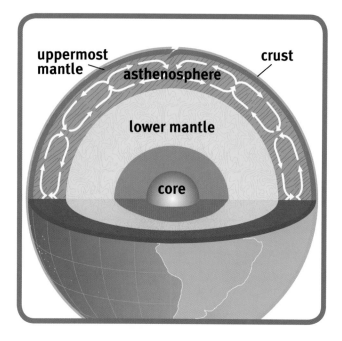

▲ Partially melted rock circulates in currents in the asthenosphere.

THE ASTHENOSPHERE

The asthenosphere is very hot. But this layer is under less pressure than the lower mantle. The rock of the asthenosphere softens and flows from place to place.

Heat rising from the core and lower mantle causes currents in the asthenosphere. Hot material rises toward the crust. Cooler rock nearer the surface sinks back down toward the core. Currents in the mantle are not like the currents in a stream of water. Mantle currents move only a few centimeters per year. It takes millions of years to travel through the asthenosphere.

The Asthenosphere

Depth:	Varies by location. Approximately 660–60 km (410–40 mi)
Thickness:	60–600 km (37–370 mi)
Temperature:	1,600°C (2,900°F)
Pressure:	133,000 atmospheres at 400 km depth
State:	Partially melted solid
Composition:	Mostly minerals containing oxygen, silicon, magnesium, and iron

(measurements rounded to the nearest ten)

THE UPPERMOST MANTLE

The uppermost mantle is cool enough to form solids. The top of the uppermost mantle connects to Earth's crust. Together these layers form the **lithosphere** (LIH-thuh-sfeer).

The mantle part of the lithosphere varies widely in depth. The uppermost mantle is thin under the ocean. This layer is thicker under the land. In some places, like on the sides of some mountains, we can even see the uppermost mantle.

The Uppermost Mantle

Depth:	Varies by location. Approximately 200–10 km (120–6 mi)
Thickness:	Up to 190 km (120 mi)
Temperature:	Under 600°C (1,100°F)
Pressure:	Thousands of atm
State:	Solid
Composition:	Mostly minerals containing oxygen, silicon, magnesium, and iron

(measurements rounded to the nearest ten)

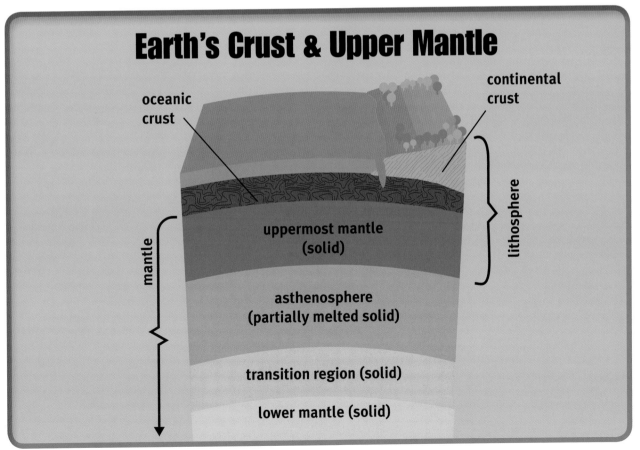

▲ The uppermost mantle and crust combine to form the lithosphere.

THE MOHO

The boundary between the uppermost mantle and the crust is called the **Moho** (MOH-hoh). The Moho is named after Andrija Mohorovicic (ahn-DRAY-uh moh-hoh-ROH-vih-sich). Mohorovicic noticed that earthquake waves suddenly sped up as they reached this depth. He thought that the earthquake waves must be entering a new layer of Earth.

The Moho lies 5–75 kilometers (3–47 miles) beneath Earth's surface. This is the depth that the Kola Superdeep Borehole project was designed to explore. The Moho is the only part of Earth below the crust that humans can hope to reach with today's technology.

The Crust

The crust is the outermost layer of Earth. Almost all life on Earth lives on the crust or in the water that pools on the crust. The crust is a thin layer. The crust is usually less than 50 kilometers (31 miles) thick. Pressure in the crust is much less than in the lower layers. The crust is Earth's coolest layer. This is because the crust is farthest from the core. The air also carries heat away from the crust. The crust is cool enough to stay solid.

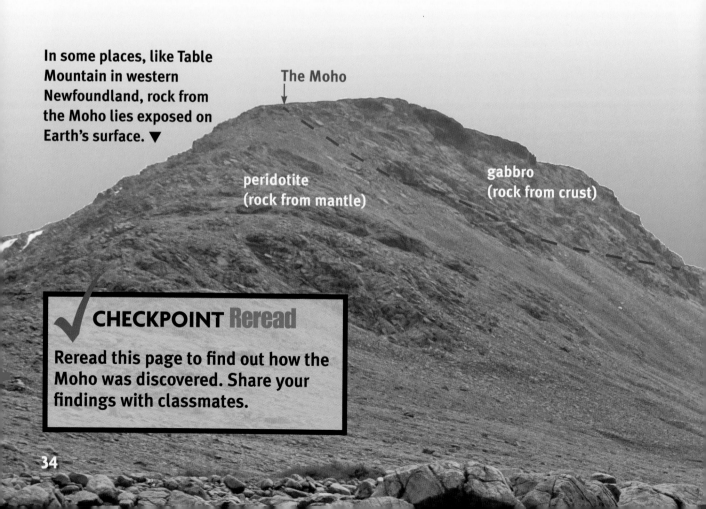

In some places, like Table Mountain in western Newfoundland, rock from the Moho lies exposed on Earth's surface. ▼

The Moho

peridotite (rock from mantle)

gabbro (rock from crust)

✓ **CHECKPOINT Reread**

Reread this page to find out how the Moho was discovered. Share your findings with classmates.

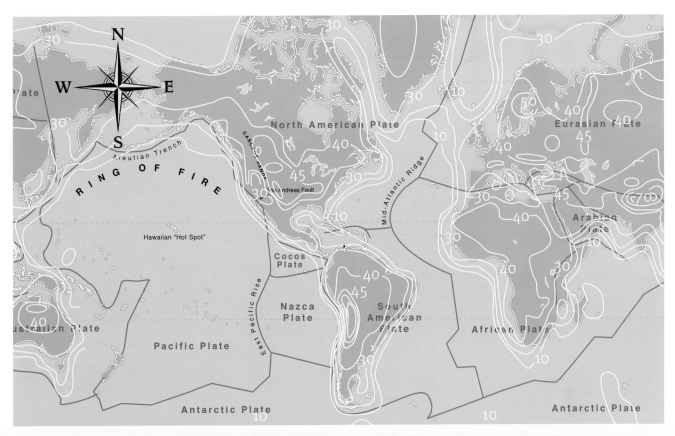

▲ The map above shows the thickness of the crust (in kilometers) all over the world. Oceanic crust lies under the oceans and is very thin and made of the rock basalt. Continental crust is thicker and made mostly of granite.

The Crust

Depth: 0 km to 100 km (60 mi)

Thickness: Up to 100 km (60 mi)

Temperature: Up to 400°C (750°F)

Pressure: Low, but increasing with depth

State: Solid

Composition: Mostly minerals containing oxygen, silicon, aluminum, and iron

(measurements rounded to the nearest ten)

Make an Earth Model

You can't visit the layers of Earth beneath the crust, but you can make a physical model of them. Just how thin is the crust compared to Earth's inner layers?

Time Required

30 minutes

Materials Needed

four colors of modeling clay or play clay, balance, rolling pin, fishing line or thin clay wire

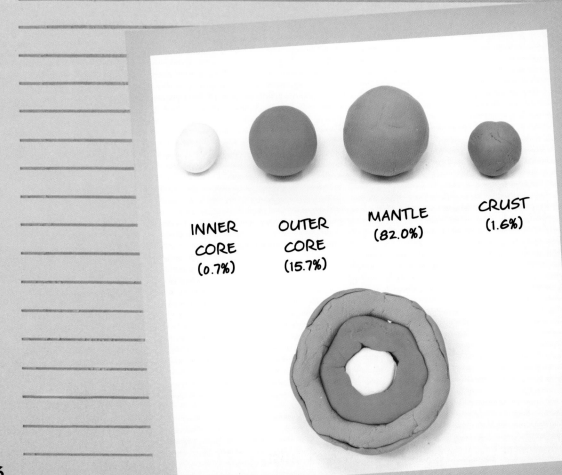

INNER
CORE
(0.7%)

OUTER
CORE
(15.7%)

MANTLE
(82.0%)

CRUST
(1.6%)

Procedure

1. Choose a color of clay to represent each major layer of Earth: inner core, outer core, mantle, and crust.

2. Your Earth model will have a mass of 300 grams. Use your balance to make lumps of clay to represent each layer: inner core, 2.1 grams; outer core, 47.1 grams; mantle, 246 grams; and crust, 4.8 grams. Roll each color of clay into a separate ball. Sketch and describe the size of each layer.

3. Roll out the ball of clay that represents the outer core. Wrap the outer core around the ball for the inner core. Repeat this rolling and wrapping with the mantle ball of clay. Roll out the crust ball as thin as you can and do your best to make it wrap around the mantle layer.

4. Use fishing line or clay wire to cut your finished Earth model in half to show the layers inside.

Analysis

1. Compare the relative volume and thickness of each of your four layers.

2. The crust has twice the volume of the inner core. Why is the crust layer thinner?

3. If the crust in your model is so thin, why is it so difficult to drill through Earth's actual crust?

FORMATION OF THE CRUST

Earth's crust formed while the planet was very young and hot. Heat from meteorites turned Earth into a molten ball of rock. The densest minerals sank and formed the core. The least dense minerals floated to the surface. These light minerals hardened into the crust. The uppermost mantle became a solid, too. These two layers float together on top of the asthenosphere.

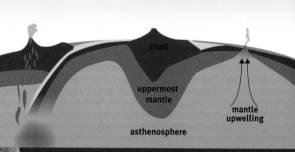

crust

uppermost mantle

mantle upwelling

asthenosphere

▲ Icebergs and continental crust both displace the material they float on.

CONTINENTAL PLATES

The lithosphere is made up of pieces called **continental plates** (kahn-tih-NEN-tul PLATES). Each large plate is made of smaller plates. The plates float on the asthenosphere. Slow currents cause the plates to move. Continental plates travel about 1 to 10 centimeters (0.5 to 4 inches) per year.

Look at the diagram below. You can see that each plate is moving in a different direction. Each year, the North American and European plates move about 2 centimeters (1 inch) away from each other. That means that London and New York are drifting a little farther apart every year!

Earth's lithosphere is made of moving plates that float on the asthenosphere. ▼

When two plates bump into or rub against each other, they release a huge amount of energy. The collision causes earthquakes.

Science & Math
Multiplying by Conversion Factors

Scientists often need to convert from one measurement unit to another. The old measurement is multiplied by a conversion factor to change it to the new units. For example, a geologist may need to convert the height of a 13,570-foot volcano to meters. The conversion factor is 0.305 meters per foot.

13,570 ft x 0.305 m/ft = 4,139 m

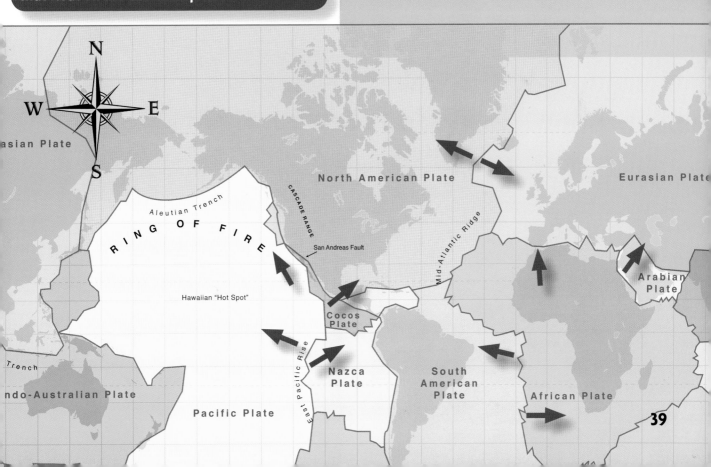

EARTH'S SURFACE

Many of Earth's surface features come from how the lithosphere and asthenosphere affect each other. Volcanoes form where hot material from the asthenosphere rises through the lithosphere. The pushing of two continental plates against each other raises mountain chains. Valleys get wider as the plates under them move apart.

You cannot see what happens deep inside Earth. But it affects you all the same.

▼ Interactions between the lithosphere and the asthenosphere create many of Earth's surface features.

✓ **CHECKPOINT**

Make Connections

Think about how continental plate movements shape Earth's crust. How would the world change if the plates started moving ten times as fast?

Summing Up

- Earth's upper mantle contains the asthenosphere and the uppermost mantle.

- The asthenosphere flows slowly as hot materials rise toward Earth's surface and cooler materials sink.

- The uppermost mantle is a solid layer that is connected to Earth's crust.

- Together, the uppermost mantle and crust form the lithosphere.

- The lithosphere consists of large continental plates that drift slowly on the asthenosphere. Many of Earth's surface features result from the interaction of the lithosphere and asthenosphere.

Putting It All Together

Choose one of the research activities below. Work independently, in pairs, or in small groups. Share your responses with your class.

1 Think of a ball-shaped food that has several layers, such as a watermelon or a hard-boiled egg. Create a poster or model that compares the layers of your food to the layers of Earth.

2 Write your own version of *A Journey to the Center of the Earth*. Use information from this book to help you describe the journey.

3 While you are studying together, a classmate tells you that *lithosphere* is another word for "crust," and *asthenosphere* is another word for "mantle." Write a response explaining why this is correct or incorrect.

The Next Step

Suppose that astronomers discovered a new planet with a strong magnetic field. What could they infer about the inner structure of the planet?

▲ Drilling near Japan provides evidence about Earth's inner structure.

The Kola Superdeep Borehole both failed and succeeded. The hole failed to reach the Moho. At 12.3 kilometers (7.6 miles), the rock became so hot that it began to melt. Every time the drill bit was lifted, rock oozed back in to fill the hole. The scientists had to give up the project. On the other hand, Kola produced a wealth of information about Earth's crust. Scientists will analyze the samples for many years.

Meanwhile, Earth scientists are trying new experiments. A team is drilling near Japan to learn about continental plate movements. Other geologists drill into Hawaiian volcanoes to study the asthenosphere. Yet another scientist proposed a dramatic experiment to explore Earth's core. He suggested setting off a nuclear blast that would open a crack in the crust. An iron probe would sink through the mantle and report on conditions all the way to the core.

Unlike the characters in *A Journey to the Center of the Earth*, scientists have not made it to Earth's center. A probe to the core may be just a dream, but scientists are learning more about Earth's structure every year. Who knows what inventions or new tools will let scientists learn more? Keep your eyes open for new discoveries about Earth's structure!

▼ **Earth's layers can be described using chemical or physical terms.**

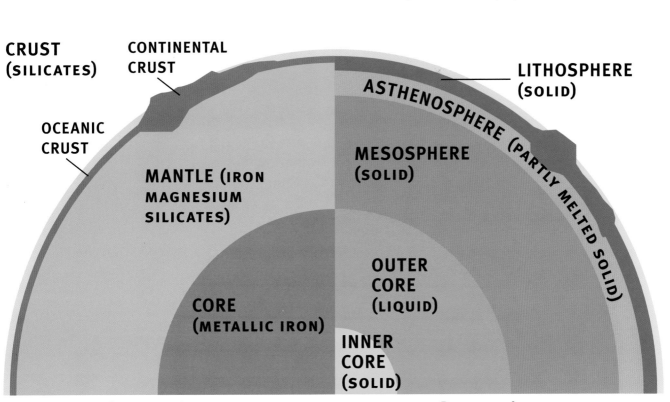

CRUST (SILICATES) CONTINENTAL CRUST LITHOSPHERE (SOLID)

ASTHENOSPHERE (PARTLY MELTED SOLID)

OCEANIC CRUST

MESOSPHERE (SOLID)

MANTLE (IRON MAGNESIUM SILICATES)

OUTER CORE (LIQUID)

CORE (METALLIC IRON)

INNER CORE (SOLID)

CHEMICAL LAYERS

PHYSICAL LAYERS

How to Write a
Hypothesis

One important step in the scientific method is to write a hypothesis (hy-PAH-theh-sis). A hypothesis is a suggested explanation for scientists' observations. Writing hypotheses helps scientists plan experiments and evaluate their results.

STEP 1

Choose an interesting observation that you don't completely understand. For example, you might notice that several recent volcanic eruptions happened when the moon was full.

▲ **Some scientists hypothesize that volcanic eruptions are more common during a full moon.**

STEP 2

Gather information by reading and observing. You could use the Internet to find news stories about volcanic eruptions. Encyclopedias and other reference books could help you understand how the moon's gravity pulls on magma inside Earth.

STEP 3

Use your information to propose one or more hypotheses. Hypotheses cannot involve the supernatural. Each hypothesis should lead to predictions that can be tested. You could hypothesize that the full moon causes an increase in volcanic activity.

STEP 4

Design an experiment to test your hypothesis. First, you might predict that there will be more eruptions when the moon is full. You could find out when the next few full moons will take place. Then you could check one of the volcano-monitoring Web sites every day to record the activity of erupting volcanoes. Record the results of your experiment in a data table.

VOLCANIC ACTIVITY IN MARCH				
Volcano	Level of Eruption			
	DAY 1	DAY 2	DAY 3–Full Moon	DAY 4
Galeras, Colombia				
Nyiragongo, Congo				
Kilauea, Hawaii				
Batu Tara, Indonesia				

STEP 5

Compare your results with the predictions based on your hypothesis. Were volcanoes more active during the full moon than other days? If the results match your predictions, the experiment supports your hypothesis. If the results contradict the hypothesis, it's time to revise your hypothesis!

Glossary

asthenosphere — (as-THEH-nuh-sfeer) *noun* fluid layer of the upper mantle (page 32)

composition — (kahm-puh-ZIH-shun) *noun* types of minerals in a rock (page 9)

continental plates — (kahn-tih-NEN-tul PLATES) *noun* interlocking pieces of the lithosphere (page 39)

core — (KOR) *noun* dense layer of iron and nickel at the center of Earth (page 6)

crust — (KRUST) *noun* solid outer layer of Earth (page 10)

density — (DEN-sih-tee) *noun* measure of how much mass is contained in a given space, or volume (page 8)

gravity — (GRA-vih-tee) *noun* force that causes all objects with mass to be attracted to one another (page 8)

lithosphere — (LIH-thuh-sfeer) *noun* solid uppermost layer of the mantle and the crust (page 33)

mantle	(MAN-tul) *noun* middle layer of Earth between the core and crust (page 16)
mesosphere	(MEH-zuh-sfeer) *noun* solid portion of the lower mantle (page 28)
meteorite	(MEE-tee-uh-rite) *noun* piece of a meteoroid that survives the fall through Earth's atmosphere and lands on the surface (page 9)
Moho	(MOH-hoh) *noun* boundary between the uppermost mantle and the crust (page 34)
primary wave	(PRY-mair-ee WAVE) *noun* earthquake wave that moves back and forth like a spring compressed and released, also called P wave (page 12)
secondary wave	(SEH-kun-dair-ee WAVE) *noun* earthquake wave that moves up and down or side to side through the ground, also called S wave (page 12)
seismograph	(SIZE-muh-graf) *noun* instrument used to record the passage of earthquake waves (page 13)

Index